The 7 traits of highly effective fire engineers

A mini book

Paul Bryant

ISBN-13: 978-1532952852
ISBN-10: 1532952856

DEDICATION

I dedicate this book to my two wonderful daughters, Ruby and Poppy. I wish them the very best in carving out a career of their own.

I also dedicate this book to all those who work in the fire industry, in whatever capacity. Our industry is unique – strange at times – but full of promise.

Paul Bryant

THE FIRE ENGINEER

In spiking high office towers,
Or deep in underground transit networks.
In heavy industrial dens,
Or in busy medical edifices.
I strive to conquer the risk,
I design to overcome the threat,
I seek for safeness for all,
I work to alleviate the fear,
For I am the fire engineer.

CONTENTS

ACKNOWLEDGMENTS

I would like to acknowledge, and thank, all those fire engineers out there who have inspired me to write this book. Some are well known in the industry. Some are largely unknown. However, all play a vital role in what makes fire engineering what it is today.

Paul Bryant

Paul Bryant

FOREWORD

This is a book about you - and about me, and about certain persons within the fire engineering profession who have that something extra - something that we can all learn from.

Now, I'm nearing the end of my career in the fire industry. I would *love* to state that, in my time of involvement in this marvellous world of fire safety, I have seen so much change.

But I can't.

I've experienced the computer develop from a simplistic piece of kit with one kilobyte of RAM, to hand held smart phones with a million times the power of my first desktop PC of the early eighties.

I've witnessed the development of internet communications from the noisy and unreliable 28.8k modems, so costly that they were only affordable in industry and commerce, to ultra-high speed fibre optic cable systems serving every home.

I've also read a recent article on the benefits of fire sprinkler systems which, amazingly, echoes the arguments in an article written back in the nineteen eighties.

So, does this mean that the fire industry is slow, colloquial or even backward?

Perhaps, anything that concerns the safety of life should be conservative in embracing new ideas? We need appropriate controls on safety regulation around the world. But do the tight reigns prevent our industry galloping ahead?

Possibly? But change is happening.

In the last twenty years or so, we have moved from fire codes that are highly prescriptive, to today's more flexible fire-safe design solutions for all types of buildings. We often describe this alternative methodology as a "*performance based approach*".

What we mean by this is, that a set of performance objectives for fire safety are set out at an early stage for each and every type of structural environment. The job of the fire engineer is to prove that, via the use of codes and analysis, that a fire strategy incorporating a fire safety engineered design solution, will meet with those performance objectives.

The turn of phrase "*there are many ways to skin a cat*" comes to mind, as it is feasible - no, probable - that any two or more groups of fire safety engineers working on the same building, may come up with alternative fire strategies. Can we have a number of design solutions that all provide for the same level of fire safety?

Well, it appears that the answer could well be an affirmative. But this then poses an interesting predicament for those who are tasked to approve the aforesaid design solution. How can one be absolutely sure that the design is fit for purpose? This is where the role of the fire engineer has to be appropriately tested.

Many in the industry can read the fire design codes and apply them.

Many can use the fire modelling and evacuation computer models, which are sometimes free downloads, to determine the two key parameters of a life safety performance approach; ASET (available safe escape *or egress* time) and RSET (required safe escape *or egress* time).

But does this mean that we will all arrive at the right answer?

Another well used term is "*trash in – trash out*". This term is highly relevant in the field of fire engineering, as the assumptions made, and the parameters set, can lead to a considerable array of results. Consequently, something as important as life safety perhaps requires appropriately high levels of diligence.

And this leads me to the point of this mini-book. Whatever we come up with is as good as those fire engineers behind that result.

My career has allowed me to meet many hundreds of personalities within our profession, throughout the world. I have met the good, the bad and the downright terrible. But I have also had the fortune to meet, and work alongside, engineers who I could easily describe as "*great*". But is the term "*great*" appropriate for an engineering discipline mostly dealing with the protection of life and property?

Perhaps the term "*effective*", or even "*highly effective*" is better? After all, the aim of every fire engineer is to provide for a strategy that is most effective in providing appropriate levels of *firesafeness* – by using the optimum level of resources, yet providing a solution which will work now and into the future.

This group of highly effective fire engineers may be found in all walks of the profession, and in any nation. They may be scientists who have developed the next generation of sprinkler heads to provide the most effective water discharge in controlling fires. They may be from an enforcement background and have helped develop robust, yet flexible, fire safety legislation.

They may have championed the cause to drastically reduce the incidents of false alarms, or may have developed the evacuation analysis modelling programmes that are used on a global basis. I would even include engineers who have been involved in sales for most of their career. Their enthusiasm for what they do, and what they offer, is also a key industry driver. And let us not forget the journalists out there who have played a vital role in the growth and understanding of the subject.

These people have all contributed to making the fire industry what it is today – in different ways. But the sum of their contributions has led to change. This is in spite of an industry perceptibly advancing at a slower pace than that which some would say is ideal. I also came to realize that these fire professionals also seem to share qualities that makes them stand apart.

So, on one hot sultry day some years back, I began my quest to find out what makes these individuals different from the rest of us. I eventually determined that there were at least seven traits that they all shared. This then led to the idea to write this book, with the aim that it may help others out there – especially the next generation of fire engineers.

Now let me make it clear that I do not consider myself one of these *greats*. However, I do believe that I have taken many lessons from them – lessons that I continue to use.

You will see, when reading the following pages, that there is nothing magical that puts these people above the rest of us. That should become steadily more obvious after every page.

We can all be highly effective fire engineers!

TRAIT NO. 1

HAVING PASSION

They love what they do....

Wouldn't it be fantastic if we could get paid, and paid well, for doing the thing that we most love to do?

Many are that fortunate, whether they are accomplished footballers, space scientists, doctors, guitarists or cooks.

But what about those who work in the fire industry?

Can we put ourselves up there with those we see on TV who seem to be paid so much for their talents? Talents that bring pleasure to the rest of us.

I do not believe that those who find themselves enjoying such a rewarding lifestyle are there by accident. They are driven by ambition. And ambition is fuelled by something more basic, and that is *passion*.

And yes, I have met a number of people in this profession of ours who do indeed display the same level of passion as the best known rock stars or celebrity TV chefs.

What drives a scientist to spend their post university career studying water droplet sizes for fire sprinkler systems?

What makes a civil engineer want to push forward the boundaries of structural fire engineering?

It is a simple and unswerving passion for what they do, day in

7

and day out. Perhaps, comparatively speaking, they may not be paid as well but financial reward is unlikely to be their main driver.

Sometimes we can forget the main reason for our profession is that we are there to ensure the safety of lives and the protection of assets against the ravages of fire. Our job is to make certain that the complex structures dreamed up by architects have the relevant safeguards built in to protect the occupants and to limit the damage caused –simply to make sure that the buildings are *firesafe*.

What could evoke more passion and satisfaction, than to be that person who provides the final fire strategy?

Yet, ask many fire engineers what they see as being their main duty and they may come back with a response along the lines of "to ensure compliance with the regulations."

If that is how we see the beginning and end of or task, how can we have passion?

Without passion, a career could be regarded as wasted. Yet, I have met many who really do not have the passion for fire engineering but have spent the whole of their working life in fire safety. And people *without* passion for what they do, are less likely to be as good a fire engineer than those who love what they do.

It is therefore obvious that the best fire engineers are driven. Monday morning is not the worst time of the week for them. It is the start of ....

more to learn; more challenges to be tackled; more to contribute.

It was when working with London Underground where I came

across the trait of passion in abundance.

Many of those in the fire safety department weren't especially qualified in the subject. Some weren't even fire engineers in the true sense but there was something obvious that they shared a real commitment to the organization - to make London Underground a safer place.

This group really loved what they did, and when they were not working, they would spend time together to talk about the mountain of issues they faced on a daily basis, and how these could be solved in the best possible manner. This was normally accompanied by a pint or two!

And it was infectious. In fact it was exciting.

Soon after the Kings' Cross fire in 1987, where thirty one souls died in horrific circumstances, legislation was hastily prepared specifically for the fire protection of sub-surface railway stations. One hundred and fifteen underground stations (or thereabouts) had to made be fully fire compliant as soon as possible. It was a mammoth task, yet was achieved generally on time and in budget.

Remember that London Underground is the oldest underground mass transit system in the world, as well as being one of the most complex. Each station is different and some are absolutely huge in layout and complexity.

Yet each station required upgrading to provide for fire detection systems, fire suppression and, most difficult of all, to utilize the highest levels of fire compartmentation.

It was mind boggling.

The sheer scale of the task would frighten off many, yet not this particular group of individuals.

Under the leadership of an extremely forceful Yorkshireman, the key players sat in one very large cold meeting room by West Kensington station, on a dark and miserable evening, to develop a strategy to ensure that every single station would meet with the new legislation, in full. After all, this is London Underground and we could not afford another major fire. Loss of confidence in the network would adversely affect travelling in London. London without efficient transport would very quickly grind to a halt. The consequences of this are just not worth thinking about.

My role was to oversee the audit of the detection, suppression and compartmentation works. This involved week after week of walking around stations during the time when they were closed to the public. This period was appropriately called *engineering hours*.

We were a small team of specialists. I covered fire detection and my colleagues covered suppression and compartmentation. We would inspect every part of every underground station between one and five in the morning. We had to check every aspect of the installation.

Auditing fire detection and suppression systems is one thing but checking that the fire compartmentation was intact, required an eye for detail, despite the unwelcoming nature of a century old infrastructure in the dead of night.

In a matter of months, every station was made fully compliant and signed off. Without the passion and commitment of this small group, there is no way we could have succeeded.

Two decades later, and some of the group are retired, or have passed way. The rest often meet up and discuss those strange but amazing days. We still have that feeling of achievement – we still have passion for what we achieved.

And passion really makes our working week one of joy.

So how passionate are you about the subject of fire engineering? It is not simply about how much you enjoy working in the profession but how good you believe you are – and how good you strive to be.

Take a look at the following table.

Without too much thought and hesitation, identify which number you believe most closely describes you.

	I don't care for the subject.	It's an OK way of making a living.	A just love the fire profession!
I am v. good at fire engineering	**1**	**2**	**3**
I am reasonably good at it!	**4**	**5**	**6**
I don't rate myself highly as a fire engineer	**7**	**8**	**9**

So what does each number mean? Check out the following assessment …

1	*Is yours a great talent wasted? Perhaps you would enjoy another profession better?*
2	*Another area of fire engineering may make better use of you, and may provide a more rewarding role.*
3	*You are a "highly effective" fire engineer - or you will be.*
4	*As with 2 – another area of fire engineering may fit better.*
5	*Are you simply sitting on the fence? Do you have aspirations?*
6	*Keep going – you will get there!*
7	*Quite definitely – wrong profession. Do not waste your life.*
8	*Perhaps you have lost ambition? Review your life plan!*
9	*Sometimes passion on its own isn't enough. Or are you terribly modest?*

And let us not forget that passion is not always a permanent state. We all have times when other priorities take precedence. There will be family crises. There will be health issues. There will be times when it is hard to have that positivity that is closely linked to passion.

But the highly effective fire engineer will ensure that temporary lapses are allowed for and that it is possible to get back to that feeling of fulfilment that comes when you do something that you love, to the best of your abilities.

Fire engineering was once described to me as a beautiful subject. If we all think of it this way, then we will start to see what the author of those words meant.

TRAIT NO. 2

BEING OBJECTIVE

They prefer to make decisions based on the facts, even if the facts contradict their personal experiences and beliefs...

Whereas many engineering disciplines are factual and based upon absolutes, I believe that fire engineering is quite different.

In the early months of my career, my boss at the time referred to fire engineering as a "*black art*". And that was the days *before* the advent of performance based fire engineered solutions. Even in those days, when all fire safety standards were highly prescriptive, there was a certain mystery behind the creation of fire safety designs.

Whereas electrical, mechanical and civil engineers use recognized and consistent laws to provide results, the determination of the most optimum means of escape, the layout of a fire detection system, or the application of a fire sprinkler system may derive from *educated guesses*, rather than hard scientific fact.

Examples include the development of escape travel distances, which had more to do with the length of a hose reel than the calculated safe speed of evacuation. The spacing of fire detection systems had more to do with one manufacturer coming up with a figure that seemed to be *about right* rather than calculating a figure based upon the dynamics of smoke movement.

As a consequence, the *black art* of fire engineering can result in solutions derived from opinion rather than fact. One engineer may come up with one solution and another with an alternative solution. Who is to say who is right or wrong? The danger is that subjectivity could potentially lead to solutions that have not been fully thought through.

This is where a highly effective fire engineer may differ from the mainstream of fire engineering. In continuing to question what they have learned to date – to always seek objectivity in their approach, they intentionally remove the influence of their personal beliefs and experiences that they may have picked up during their career.

Let me provide you with an example. I was involved in a research project to assess how occupants of railway stations react to fire signage. The project developed from an initial assessment of how the average person is aware of emergency signage.

The project then progressed towards the creation of active dynamic signage systems that would guide persons to the safest routes. These were again tested by live groups of people and new lessons were learned which would add to our knowledge bank of evacuation philosophies and systems.

The project was led by an experienced and reputable scientist and engineer who had spent his lifetime studying how people evacuate from buildings, railway carriages, aircraft and ocean going vessels. There was little that this person did *not* know about how we react in an emergency. Yet, this person intentionally kept an open mind.

As we moved through the project, so many of the results confirmed previous conclusions from the projects of the past.

Yet this fire engineer continued to ensure that his team remained objective – that they never assumed that they would know the answer before the hypothesis was properly tested.

Fortunately, this open minded neutrality proved to be correct. Results of the research led to a new number of conclusions. Simple factors that may have been overlooked now proved to be influential in making sure people do the right thing and go the right way in an emergency.

Consequently, by remaining objective, we have further advanced our understanding of workable evacuation strategies.

I've encountered varying degrees of objectivity, and subjectivity, throughout my career.

I remember a long time ago when I was in my twenties, being asked to audit a major fire detection system utilizing computer graphics as the main form of indication. This was at a time when the PC was in its relative infancy, although there was already a desire to see fire detection zones on a computer generated graphic display.

The designer and installer was present with the client's representative (an international airport) whilst I undertook the inspection.

I found that, by simply removing an electrical plug from its socket, I could completely eradicate the indication display. There was no back-up, no alternative power supply. There were no alternative default displays. The reboot took over ten minutes. Obviously the system did not conform to the relevant British Standards of the time.

In response to the findings, the designer / installer said to me...

"I've been in this game for over 20 years. Who are you to tell me?"

It was quickly apparent to all that this man had thought and believed that there was nothing more that he had to learn.

The lesson here is something that we can all be prone to. By continuing down the path of self-belief without, at least occasionally, pulling ourselves up, we could end up making mistakes.

Not so with highly effective fire engineers. They will continuously monitor their own working methods and understand that they may unintentionally revert to ways that could impact on their objectivity.

So how objective do you consider yourself to be? How would you respond to the following statements?

1. I've used the fire code so often, I never see the need to revisit it.

2. I never bother to read the latest news and views on fire engineering. It is always more of the same.

3. These new alternative fire suppression products will

never be as good as sprinkler systems.

4. I can't see performance based fire engineering solutions ever replacing national building regulations.

5. The problem with architects is that they never take the time or trouble to understand the fire safety issues.

6. Why try to include for building protection and business continuity in a fire strategy? All our clients care about is achieving compliance.

Be honest with yourself. If any of these statements rings true in your own thoughts, then you have allowed personal judgement, no doubt based upon your experiences, cloud your view.

As with any of the traits listed in this book, being human does mean that none of us can stay completely objective on a continuous basis. There will be times when we allow our beliefs to take over. But the lesson here is to always aim to embrace alternative solutions, even if they do not initially align with what you think.

TRAIT NO. 3

KNOW YOUR LIMITS

They know what they know but, <u>more importantly,</u> they know what they <u>don't</u> know...

Means of escape calculations / ASET and RSET analysis / structural fire engineering / point type fire detection and alarm systems / aspirating smoke detection systems / linear heat detection systems / flame detection systems / video smoke detection systems / mechanical smoke control systems / AOV systems / voice alarm and PA systems / fire and evacuation modelling / fire sprinkler systems / gaseous fire extinguishing systems / condensed aerosol suppression systems / emergency lighting and signage / firefighting arrangements / performance based assessments / the psychology of human behaviour in an emergency / fire compartmentation / fire doors / fire damper systems / door hold open devices / fire and smoke curtains / dry and wet riser systems / intumescent coating systems / flammability of materials assessments / fire investigations / control of ignition sources / the science of fire growth / fire risk assessment / fire strategy preparation / portable fire extinguishers and blankets / hose reel systems / firefighting techniques / foam application systems...

The above list is not exhaustive but provides a list of topics covered by the singular discipline of fire engineering.

We, in the fire industry, understand how multifaceted the subject is. The outside world does not.

Imagine for a moment that you are in a high level meeting with various stakeholders. All the key people are there, including the architect, project manager and the relevant enforcement agencies. A question is raised about one detailed feature of a fire engineering based problem. All heads turn to you They will expect you to be able to offer a complete and through response. After all, you *are* the fire engineer.

However, it is a specialist area of fire engineering outside of your expertise. You do not feel that you can give a proper authoritative answer.

Most of us have experienced this at some stage in our careers. If faced with a question that we do not feel we can answer fully and accurately, we will be forced into one of two responses.

Response 1 is to admit that you do not know the answer.

Response 2 is to provide an answer that may not be correct.

Now let's be honest with ourselves for a moment. If we do go with Response 1, it could lead to embarrassment all round. Others in the meeting may subliminally at least question your professionalism and wonder why you are at the meeting.

Response 2 is an easier option to live with short term. If the answer is subsequently proved to be right, then all is good with the world. If wrong, then we have bought time to correct our mistake. We could blame the incorrect answer on not fully understanding the question, or put it down to aspects out of your control – "*oh, the standard has changed*!"

Unfortunately, the competitive nature of the fire industry is such that there is a need to protect the good name of both ourselves and our organisation. *Not knowing* is seen as defeatist. We cannot let ourselves or our business down. A theme that you will find again in this book.

And that is the sad truth.

Not knowing is often regarded as an admission of ignorance, especially in a world where the art of fire engineering is not trusted by many outside of the industry. So can we ever admit to not knowing an answer to a question that may seem to be easy for those not in the know?

My experience has shown that admitting that you do not know isn't really a problem if you handle it the right way. What is wrong with...

"I'm sorry. May I investigate before I provide a proper response to that question?"

But it is even better to plan properly and know when, and how, to respond to any fire engineering issue. Those people who I describe as highly effective would normally surround themselves with persons with a range of expertise and experience. If they do not know the answer, they sure as hell will know of someone who does.

And that is the key message. To understand that none of us, even those who could be described as experts, can know everything there is to do with fire safety and engineering. Not knowing is not ignorance but an admission of being human.

This profession is full of high levels of expertise, with many concentrating on one facet of the subject...

Specialists studying the efficacy of sprinkler systems.

Engineers who have developed structural fire safety.

Computer programmers who offer fire modelling capability utilizing the equations of fire science.

Manufacturers who have invented new intumescent coating materials

Ask them about fire detection, or smoke control, and they will likely be lost.

This is what makes the fire safety industry so unique – so interesting. It is so diverse. Collectively as an industry we can provide an answer to any fire engineering query. It really is a profession where teamwork is key, and those highly effective fire engineers have always realized that.

Now a question for you, the reader. Have you ever properly examined the scope of your knowledge? What areas are you an expert in? What areas leave you cold?

In the following exercise, I provide a simple method to help you assess how much you really do understand. Over the page is a simple table using three columns.

The first column will be a list of the subjects you feel you understand very well. The second lists those areas that you have a good basic grasp on.

The third will be areas that you have little knowledge of. In order to understand all the subjects contained within fire safety and protection, you could use the list of subjects given at the beginning of this chapter.

You may produce something like the following:

EXPERT	A WORKING KNOWLEDGE	LITTLE KNOWLEDGE
Fire detection systems	Passive fire protection	Fire modelling
Voice alarm systems	Means of escape calcs.	Structural fire engineering
Smoke control	Sprinkler systems	RSET/ASET calcs.

Now review your answers. In the above case, an initial analysis would assume that this person came from a systems background and possibly moved into fire engineering. What does your response tell you about you? Do you see it as an *aide mémoire* to develop the areas you are less knowledgeable in, or do you see the list as a *fait accomplis*?

By being true to yourself, you can finally make tangible your knowledge strengths and gaps. If you believe that there are no areas that you are an expert in, do not despair. Possibly you are being too tough on yourself. Alternatively, there is nothing wrong with being a generalist, as long as you understand this, and are able to find the expertise when required.

I would recommend that you undertake this task at least twice, say a week or more apart - just to ensure consistency.

If you are part of a team of fire engineers, then it would be interesting to pool your results and see how, collectively, you have command over the various facets of the profession. A team that can show expertise in many areas can be extremely powerful.

Paul Bryant

TRAIT NO. 4

PROGRESSIVE THINKING

They look to the future – they don't live in the past ...

I remember when I was involved in standards making, back in the good old 1980's. I was attending a British Standard meeting covering a specification for fire detection control and indicating equipment. Computer technology was still in its relative infancy yet there was a move to use microprocessors for fire detection and alarm equipment.

Some members of the committee were not happy with this. In their minds, microprocessor technology was highly unreliable. Their experience was of home or office computer systems that often crashed. As we are dealing with life safety, systems that failed were obviously not welcomed.

I clearly remember the phrase along the lines of "*not in my lifetime will I accept microprocessors to be used in fire systems*". As a result, a team of experts drafted an annex to the standard, describing how the architecture of the microprocessor should be designed. In hindsight, how inappropriate – how parochial, but that was then.

But have we really changed?

I highlighted in the introduction to this book that we are still advocating the benefits of sprinkler systems. This is much the same as we advocated those same benefits some thirty plus years ago.

I have always been a big fan of fire sprinkler systems, mostly because they work! In my mind, automatic fire sprinklers would benefit every type of building. Yet we still seem to have a hard time in gaining common recognition of the benefits, particularly outside of the United States. Is it due to the clients not wanting to pay an extra amount for such protection or is it the fault of us, as collective fire engineers, not being strong enough in our representations?

After consideration, I do believe that at least some of the fault is with us. There are so many new fire safety and firefighting technologies out there that work, yet I know of certification bodies who will not automatically recognize such advances because there is "*no money*" in creating schemes for only a mere handful of manufacturers of new technologies

And that goes for fire safety engineering. Proprietary fire and evacuation modelling programmes have become so advanced that this is really the only way of properly determining ASET and RSET for new buildings. Yet there are still quite a few fire engineers in senior positions who refuse to acknowledge the benefits of modelling, and continue to use calculation methodologies that are simply well out of date.

The fact that fire safety is a subset of life safety has created a high degree of risk aversion. But that is not the real reason. I think that there are many who simply do not advocate new ways because they do not understand them.

As an exercise, let us look at all the range of fire safety and protection systems and technologies currently on offer. Some are tried and trusted mature technologies that are never in question. Some are the latest technologies that have not been fully taken up as yet but show promise. Some are technologies that have been introduced yet are now regarded as largely redundant.

So, using your judgement, enter the technologies that you personally believe fit into each of the three boxes:

Fire Systems Technologies

Mature	New	Redundant

As with the last exercise, you can use the list of subjects at the beginning of the chapter covering trait number 3 to assist but ideally you will use your own knowledge of the various technologies out there.

By carrying out this task, it forces you to think again, which has to be a good thing. Now let us move on.

British Standard Specification PAS 911, covering fire strategies, was published in 2007. In this document, it included a method to aid objectivity (see Trait number 2) in choosing appropriate systems for specific tasks. I named the table the "*quantified assessment of options*" and based the evaluation criteria on three factors;

Performance – how a system will perform within a specific

environment.

Logistics – the practicality of installing and maintaining the system within that environment.

Economics – the whole life cost of the system within that environment.

The idea is that each technology option is scored against these three factors. Each score is then multiplied and a total score is given.

Given that some projects may favour one factor more than another, you may weight them accordingly, i.e. the most important criteria is given the maximum possible score, the second the next level and the third, the lowest possible score.

Criteria	Option A	Option B	Option C
Performance (Score from x)	N1 out of x	N2 out of x	N3 out of x
	Multiply by	Multiply by	Multiply by
Logistics (Score from y)	N4 out of y	N5 out of y	N6 out of y
	Multiply by	Multiply by	Multiply by
Economics (Score from z)	N7 out of z	N8 out of z	N9 out of z
Total Score			

Taken from BS PAS 911 i-copyright Kingfell 2007

The purpose of the table is to reduce the level of subjectivity in deciding the best system type for a specific application. An example of how such a table can be used is given below.

In this case, it is comparing three different methods of smoke detection system for use in a building of historical importance. Note that, in this example, the scorer has determined that *performance* is the most important attribute, followed by *logistics* and then *economics*, based upon the maximum score setting for each. In this case, the exercise has deduced a wireless point smoke detection system is deemed the most appropriate.

Note that, had the weighting shifted, it is possible that a different system technology may have scored highest.

Criteria	Wired point type smoke detection	Wireless point type smoke detection	CCTV based point smoke detection
Performance *(Score from 10)*	8	7	7
	Multiply by	*Multiply by*	*Multiply by*
Logistics *(Score from 7)*	2	5	4
	Multiply by	*Multiply by*	*Multiply by*
Economics *(Score from 6)*	4	5	4
Total Score	64	175	112

Taken from BS PAS 911 ii - copyright Kingfell 2007

Let us go back to the table where we listed mature, new and redundant technologies. Take two or three similar system types from two or all columns from your list.

Now use a current project you are working on. Consider one of the key environments. Also consider how you would assign a weighting for each the three factors in this case. By using the table, see what of the two or three technologies chosen, scores highest.

The purpose of this exercise is more to do with evaluating technologies based on their pure merits. It is to avoid pre-judgement. It is to promote an alternative way of viewing the options we really do have. Even technologies that we may regard as obsolete may still have merits in specific circumstances.

Always be open to new concepts. And, at the same time, be prepared to properly assess the merits of those concepts – their strengths and weaknesses. In this way, every new technology that appears will not simply be written off without due cause.

Highly effective fire engineers are naturally progressive thinkers. They will consider new technologies and ideas, and will measure them against their own rigorous benchmarks.

TRAIT NO. 5

TAKING RESPONSIBILITY

They automatically assume responsibility for their output. They readily own up to mistakes ...

How many of us have made mistakes?

It is a rhetoric question of course. Mistakes are as human as everything else in our lives. However, a mistake in the profession of fire safety engineering could be potentially disastrous. If we miscalculated the ASET in a busy public interchange on a new railway station, then it is not inconceivable that there is a chance that the evacuation may not go as planned – that people may perish.

Given that mistakes are inevitable, and the consequence of mistakes are potentially disastrous, what can we do about the situation?

Well let's take a look at our options:

Option one is to deny that you had made the mistake. It could be blamed upon lack of information or being provided with inaccurate information. It could be put at the hands of others who had described the situation differently. But at the end of the day, the mistake was not due to your inability but due to other factors outside of your control.

Or the other option is to own up to the mistake. Not a great place to be in I admit. It could result in more expense, more time delays and a falling off of trust in you and your organization. But ultimately it has to be the right option given that the hiding of mistakes, especially in the realms of fire engineering, could have far greater consequences down the line.

To be responsible, and to take that responsibility properly, is what we as fire professionals strive for. Obviously we would love to be acknowledged as being responsible for a successful project conclusion. We all enjoy being praised for what we have achieved but there will be occasions in our career where things will go wrong.

Those who I would describe as highly effective, will take their responsibilities seriously yet, at the same time, will take measures to ensure that mistakes are controlled.

For a start, *they* will make sure that they understand the subject matter properly. Highly effective fire engineers will rarely bluff their way through a project.

They will read through all relevant standards and supporting literature – to get to properly understand all pertinent issues.

They would talk to others who are more knowledgeable than themselves on certain subjects.

And, as highlighted for Trait no. 3, they will normally surround themselves with people who are good at what they do.

But when it does go wrong, not only will they accept

responsibility for themselves but also for the people around them.

Taking responsibility for mistakes is an honourable thing but I have also seen it taken too far. I have met fire engineers who own up to their mistakes – time and time again. They will be the first to put up their hands – frequently and regularly.

This leads to another thought. Perhaps there are occasions when those persons making repeated mistakes are not in the right profession?

But there is another area of responsibility that I believe is commonplace in the fire industry, and that is knowing who exactly is responsible for a task. Let me give one example of this.

My career in fire has often introduced me to situations of conflict, which often revolve around the subject of responsibility. Perhaps many reading this have viewed reports or strategies that provide something I will describe as an "indicative design". This may be supported by a phrase along the lines of "The actual design is the responsibility of the fire systems contractor".

In my career I have come across many cases such as this – when somebody has been paid to provide a fire protection system design, yet hang the ultimate responsibility of that design on the poor old installation company.

Fortunately this situation is becoming less common thanks to certain highly effective fire engineers changing the requirements, so that a designer has to be explicitly stated on certification forms. But this highlights that responsibility may not be clear cut.

Another issue I have found to be relatively common, is a certain confusion between the terms responsibility and authority. Life itself seems to enjoy playing on this point.

Many of us have been in positions of responsibility yet precious little power to take the appropriate actions to ensure that we can properly assume those responsibilities. And conversely, some of us may have had the authority over certain tasks but have assigned responsibility to others. This is sometimes attributed to age and seniority.

As we march through our career, if we are *lucky*, we may get to a point where we attain the power and the privilege that goes with seniority but not the downside – and that is often *responsibility*, which is passed to the lower ranks.

And why not?

We have had to endure those years of being the fall guy.

We have had to take the blame when others have got it wrong.

*We have had to bite our lip when our seniors (our "betters")
make us do something that we know will end in failure.*

So why not, when our time comes, continue in this vein? Why not enjoy the fruits of our career climb?

As an exercise, where do you think you currently sit in the following matrix?

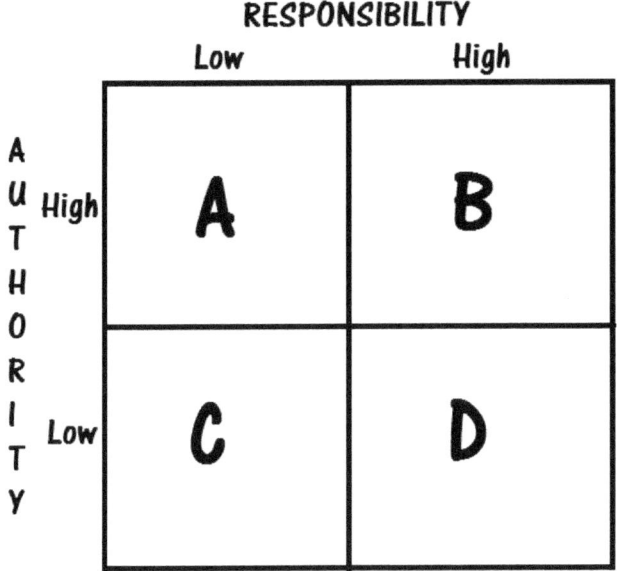

If you are in *A*, you could say that you are fortunate but this is not a place where a highly effective fire engineer would want to end up. They would tend to gravitate to *B*. C is a place where most of us start at, whilst D is a place where none of us would truly want to be, although it is where many of us tend to find ourselves at one or more stages in our career.

But what if we could choose where we would like to be – today? Many of us would opt for B, but there may be many more who would go for A.

The problem with low responsibility and high authority is that the fire industry can never really move forward. It is all too easy to maintain a status quo. After all, there are often no prizes for one to put their head over the parapet – so to speak. And the highly effective fire engineer knows this. The best of the bunch will adopt philosophies along the following lines:

1. Where those are assigned with responsibility, always make sure that they are also given the appropriate level of authority to make decisions.

2. Allow those who have been assigned with responsibility to make mistakes but, at the same time, have systems or processes in place, to make sure that those mistakes do not lead to adverse final decisions.

3. Never to use authority to dominate others.

4. Understand that those with authority have a responsibility to use that authority wisely.

If we all follow these rules of life through our career, then undoubtedly, it will be a fairer and more rewarding profession to be in – for all.

TRAIT NO. 6

GETTING INVOLVED

They really get involved in the industry ...

Even if our industry is not steaming ahead, new ideas and techniques are always being introduced. Perhaps not at the pace of other more high tech industries but there is a continual need for fire professionals to ensure that they are aware of what is going on.

Traditionally, we could regularly update our knowledge by reading fire journals, or by fire safety group meetings or conferences. However, social media has now taken over as the primary source for information. There are distinct advantages in the use of social media, primarily that it is a global platform and provides us not just with an appreciation of issues that affect our national fire industry but widens our scope to understand issues affecting other nations.

We now have the ability to understand specific problems faced by fire engineers around the world, and provide responses based upon our own knowledge. Furthermore, we can read the responses by others. This inevitably helps us all, by providing assurance that our knowledge is up to date.

Then of course there are forums where we can take part in specific subjects. We can deliberate with others in real time. There is little doubt that social media is helping the collective

Paul Bryant

conscience of the fire industry.

But this is not what I refer to when I consider involvement. In the industry and when considering the role of the highly effective fire engineer, I am thinking morseo of putting something back into the profession, and often without expecting personal gain.

Perhaps the easiest way of describing the trait of involvement is to give example of where, and how, we as fire engineers can be involved.

Trade shows

Fire industry trade shows are not just a vehicle for selling products and services but are a vital opportunity to meet like-minded persons. They provide a hub for fire industry professionals to discuss everything that impacts on their latest work role or possibly their career. There are many fire trade shows around the world and there are a hard core of personalities who attend many of these events.

Even if you have nothing to sell, or nobody specifically to meet, do not believe that trade shows offer no value for yourselves.

For a start, it is worth just being aware of the latest technologies in active and passive fire protection. I have often come away with a number of lessons learned that I was able to use in subsequent projects.

Even the *après show* events, where we can all meet around the bar with other fire engineers from around the world, are highly valuable in improving our knowledge of the technologies and the latest thoughts within a profession that benefits from new ideas.

Conferences

As with trade shows, conferences covering one of more aspects of fire safety and protection, are popping up all over the world. Whereas, once upon a time, fire safety was regarded as a singular and all-encompassing subject, there is an increasing trend to provide conferences that specialise. We now see examples of events ranging from emergency evacuation of tall buildings, to fire safety in healthcare premises, to fire protection of tunnels. Every specialist area attracts its own group of conference goers.

The fact that this is happening is extremely promising, in that it is now recognized that the application of fire safety and protection is much more complex than many outside of fire may appreciate.

Highly effective engineers will not just attend conferences but are often likely to be presenting papers. This is a natural next step for any of us who want to move up the ranks. It is almost a rite of passage to stand up in front of fellow fire professionals to present your thoughts and ideas.

I still remember my first presentation, when I was in my mid-twenties. Despite the nerves, I was able to deliver twenty or so minutes on the subject of a British Standard covering fire detection and alarm systems. I was employed by a UK fire insurance organization. A key early role was to ensure that property protection was included in the British Standard. This led to the property protection (P) classification of BS 5839 Part

1. The experience was ultimately rewarding. I enjoyed it so much that I wanted to be up there on stage again. Within a year I had presented papers in France, Germany and as far afield as Ecuador.

Of course public speaking is not for everyone, but it most definitely is one way of increasing your involvement in the industry.

Fire associations

They may be referred to as associations, institutions or chapters but there are many bodies where one can become members and participate in meetings and events.

Some associations date back decades and there was a time before social media where they were the primary focus for people to learn about aspects of the industry.

In nearly all cases, those who I regard as highly effective are members of one or more associations. They may be there to learn or to participate but I do think that those who are seeking to grow within the profession should seriously consider being involved in one way or another.

Standards Committees, Working Groups and Panels

Have you ever wanted to be involved in the writing of a national or even an international fire safety standard? And what about being on the inside when it comes to putting together the latest piece of fire safety legislation?

Many may believe that they cannot be involved in the making

of standards that affect what they do but you may just find that it is possible, if you truly wish to be involved.

The chance of sitting at a committee, representing yourself, is unlikely. Although, if you are a member of an association or institution, it may just be possible that you can be nominated by that body to sit on such meetings.

But please do not expect to be paid to be involved in standards making. In most cases it is on a voluntary basis and the time spent will often be your own time. Nevertheless, direct involvement in standards making is an experience like no other.

By being involved, one can experience first-hand how rules are drawn up. You will understand that, behind every written clause, there is a viewpoint (and quite possibly a counter viewpoint). I have spent thirty years in standards making and despite many hours passing when you would wish you were somewhere else, there were times when some of the discussions and views were quite fascinating.

Journal articles and white papers

Writing an article for a fire publication can be time consuming but always rewarding when you finally see your words in print. Many fire journals are always on the lookout for interesting articles. This is not only a great way of getting your thoughts out into a wider readership but every article can add something to the knowledge database of our profession.

And if you have some serious new ideas of how to change the industry, then why not write a white paper. You my just find that the process of writing will not only result in something new for discussion but it can help focus your own thoughts and beliefs. You may just find that by undertaking a task such as this will help create a better you.

Write a fire book

And why not? Yes it requires time and dedication – as well as a need to have something to say. But every new book on the subject will provide another perspective – making our cumulative knowledge ever more fruitful!

Involvement in the industry is something each of us can do more of. We may use excuses such as lack of time but when we do, we will ultimately find the act of involvement good for our career and, moreso, good for our soul.

TRAIT NO. 7

APPRECIATING OTHERS

They acknowledge and appreciate the skills of others in the profession. They understand that it does not have to be a competition ...

There is a theme emanating from this book that the fire industry has created a high degree of competition between those who would describe themselves as experts. There is no quicker way to gain in a mature industry than to put down a rival organization. And that is what has been happening over decades.

Check the websites of many fire consultants and they concentrate on how clever they are and how many wonderful projects they have been involved in. What they of course will not show is how effective they have been in delivering their services. This situation is endemic in many industries but is undeniably of concern in the industry in which we participate in.

If we look at other mature or declining markets, similar strategies are observed. Rather than mutually developing the market, the easier alternative is to fight for the existing and precious market share.

Many compare the fire industry to the domestic insurance industry – a grudge purchase that is mostly necessary but where the consumers use up their energies searching for the lowest price. Unfortunately this does not help the industry over the longer term. In fact, if we look at the insurance industry, the real winners appear to be the web domains who help customers choose the cheapest product.

Typical of any sector of business that finds itself in this position is to become inwardly focused. A limited market and the inability to increase it greatly, creates intense competition. Intense competition creates mistrust and the need to be seen as better than your competitors to maintain or increase business. This could be seen as a very worrying trend that could harm the industry forever. It is not a recipe for long term success.

I believe that we can all agree that a key industry driver is the need to comply with national fire safety legislation.

I remember in the early years of the new millennium, a total revamp of the United Kingdom fire safety legislation was proposed. The cornerstone was the introduction of the requirement that every building covered by the legislation would be required to have an up to date fire risk assessment. It was believed that this would lead to a fire consultant's charter. That the influx of this massive work stream would make fire consultants rich. Some were already planning their next Ferrari! (Okay, so I stretch the point a little).

But the exact reverse happened. As the legislation took hold in the mid part of the noughties, those offering fire risk assessments came from anywhere and everywhere – so long as they had a background in fire. A good number were not appropriately qualified in undertaking such assessments but that did not seem to be a hurdle. Very quickly there were many offering their services at cut down prices and customers were quite happy to pay nominal sums to have that piece of paper stating "fire risk assessment". With a market ready and willing to offer what is a professional service at a cut down price, the "responsible person" (someone with responsibility for fire safety within an organization), could pick and choose, and often went for the lowest price.

I remember writing a fire journal article in 2010 – four years after the UK's Regulatory Reform (Fire Safety) Order 2005, was put into practice. It was titled "The dangers of cheap and cheerful fire risk assessments". It seemed to stir up a lot of interest. In fact, I had a number of requests to re-publish it in various guises – one request came from the equestrian sector.

Of course, the UK fire industry was not properly prepared for this. Third party certification schemes were introduced. The term "suitable and sufficient" was bandied about in an effort to point out the fact that there are, conversely, "unsuitable and insufficient" assessments being undertaken all over the country.

Six years on from my article and I am not sure how much has changed. There have been a number of cases of litigation against both fire risk assessors and those employing them. It is also still recognised that not every building requiring an assessment has one in place.

The perceived value of a professional fire risk assessment has also been dented by the way they are sold. I understand that a fire risk assessors' hourly rate is still not up there with a plumber or electrician.

And as a consequence, fire engineering is not seen as vastly a different function than fire risk assessments to the world outside of ours. So, whether we like it or not, price continues to be a key determinant when a client chooses the services of a fire engineer.

But this need not inevitably be the case going forward.

Our sister discipline – architecture, does not suffer from this dilemma. After all, who would seek the cheapest architect to design the next grand building?

I hope that I have not painted a dark picture of the failings of an entire global industry. We must also realise and appreciate that there are clients who really do understand the benefits of good quality fire engineers. These tend to be people who are responsible for larger and more complex infrastructures – where a fire can result in a devastating outcome. Perhaps it's just that we need to persuade more to see it this way?

As an exercise, please look at the diagram on the following page. It is simple enough. The left to right measures "Market share" i.e. how much of the business, in the sector chosen, an organization has – normally measured in percentage. The down to up coordinate measures "Skill set" – i.e. how much an organization has the skills and experience to fully undertake its tasks without outside assistance. This is an arbitrary measurement from low to high.

Now choose, say, six direct competitor organizations, preferably serving within the geographical sector you are working in.

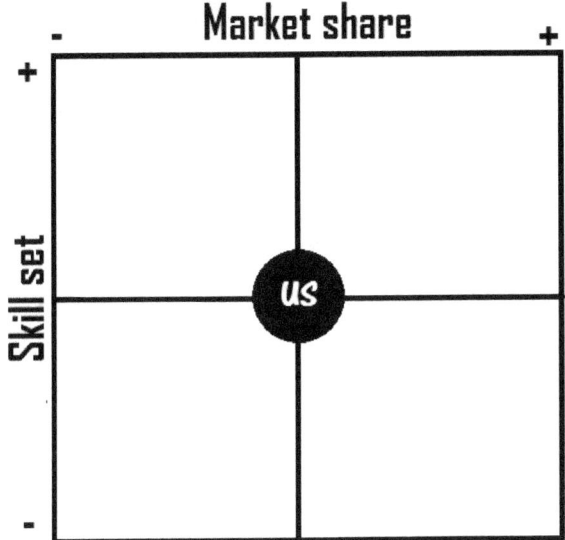

Without thinking too deeply, plot each of the organizations on the diagram. Note that your organization sits dead centre. For example, if one of your competitors has a relatively small share of the local fire engineering market, but does have a team of highly specialized engineers, then you would plot this as a point somewhere in the upper left quadrant.

Now leave the completed diagram for a while. Go and have a coffee.

On your return, review why you have entered each organization where you have. Were your initial decisions based upon gut feeling or was there any factors that would support your decisions?

Next, take your time and mark your points again. Have you made any changes to the original positions?

Let us now, hypothetically speaking, imagine that a major project requires you to collaborate with your competitors and that you all pool resources. Where would you plot this combined team as a point on the diagram? Is it an overall improvement on each of the competitor organizations?

And that is the point of this exercise. Our combined skill set will improve and together we can enjoy a better share of the market. Trait number 3 recognises that we all cannot be specialists in every aspect of the world of fire engineering. The subject is just too vast to understand the intricate details in the application of every type of fire protection system and their use within a detailed fire strategy. Many of us do have expertise in one or more of the subject areas and a general knowledge of other areas. And this is not just at an individual level but also within organisations, large and small.

But a professional future for fire engineering could look at how we as professionals collaborate and not just compete.

There is evidence that more and more fire safety engineering specialists are willing to acknowledge the skills of others. They are willing to work alongside those who have traditionally been regarded as *the competition*.

It is recognized that only the larger consultancies' can employ a large team of engineers in order to have an intricate ability to cover every aspect of fire safety and protection. There are also a good number of *boutique* fire engineering consultancy practices - small companies who offer specialist services in specific aspects of fire safety, protection and engineering.

Instead, recognition and appreciation of others in separate organisations is where we need to be. And this has already been practiced, day in and day out, by those I would describe as highly effective.

Rather than keep projects to themselves, they may choose to bring in associates, possibly from other organisations in order to provide the level of knowledge that will deliver the right result. They will not simply try and *make it up as they go along* in order to keep a project to themselves. They realise that the selfish approach will quite definitely damage the profession in the long run.

I have always been an advocate of peer review – using others not directly involved in a project to check and comment on work that we have produced. I would find a certain comfort in allowing a fellow professional to assess my work with experienced but fresh eyes. Given that fire engineering will continue to become more sophisticated and that we will be dealing with increasingly more complex projects, is this not the way to go?

We really do need to appreciate the abilities of others. Collectively we can solve any fire engineering problem. Collectively we can avoid the mistakes of the past that may have been noticed at an earlier stage if peer review was part of our process.

Highly effective fire engineers know this and will continue to see the betterment of the profession as a priority over personal short or medium term gain.

CONCLUSIONS

To be a highly effective fire engineer, you need to have good qualifications?

To be a highly effective fire engineer you need to be highly experienced?

Both statements could be correct – but that is not the message of this book.

Each of the seven traits are more to do with attitude than anything else. It is attitude – and the right attitude, that separates those who succeed from the rest. The term "*glass half full*" is probably too simplistic to describe the qualities of a highly effective fire engineer. It is more than a positive attitude but an understanding and acceptance of their professional role. It is more to do with a *want* and a *need* to excel in what they do.

Life is really too short to pursue a career in something that does not spark you in the morning. That buzz one gets as a project successfully comes to a close is quite palpable in the body language of a highly effective fire engineer. And their buzz can be highly effective and infectious in getting others on board. Not just others in the profession but those who work alongside the fire engineer – the architect, the enforcer, the project manager, and so on.

I have watched the most cynical of stakeholders being won over by these fire engineers. It is a joy to see and really helps to advance the cause of a profession that so many of us really do care about.

But then again, the traits listed in this book are not just specific to the fire safety industry but to the majority of professions out there. The traits can otherwise be referred to as *success factors* – applicable to all walks of life.

But this profession is now in the hands of the next generation of fire engineers – those who have been brought up in a world of powerful fire and evacuation modelling systems; the latest in fire protection technologies; wider acceptance of performance based fire engineering; and of course much more dynamic and complex structures to which they can apply their talents.

For those aspiring to be highly effective fire engineers, don't just decide what sort of fire engineer you want to become but what sort of person you want to be.

I truly believe that the future of the profession is in safe hands.

If you liked this book and would like to learn more about fire strategies, why not read "Fire strategies – strategic thinking" by Paul Bryant.

Available on Amazon.

Check out www.kingfell.com for all our titles.

Paul Bryant

ABOUT THE AUTHOR

Paul Bryant worked in the fire insurance and certification industry for many years before taking the role of Head of Fire Engineering at London Underground Limited in the 1990s. He later formed Kingfell which grew into a multi-million pound organization covering both fire consultancy and systems engineering services.

Paul has always been involved in writing since his early career. He has helped write many British Standards and authored British Standard Specification PAS 911, covering fire strategies, in 2007. He currently spends his time writing under the Kingfell Publications brand, speaking at conferences, and providing fire strategy services via his partnership organization, Fire Cubed LLP.

Paul is a UK chartered fire and electrical engineer and a liveried member of the Worshipful Company of Firefighters (London).